Refresher 1

Refresher 1

by Gordon L. Ziegler

Author:
Gordon L. Ziegler
P.O. Box 1162
Olympia, WA 98507-1162 USA
ben_ent100@msn.com

To order additional copies of this book, contact:
Xlibris LLC
1-888-795-4274
www.Xlibris.com
Orders@Xlibris.com
542560

PREFACE

The most important invention in human history is the Refresher 1. If it works as calculated, it would reverse all aging, disease, and decay processes in humans, animals, vegetation, and minerals in its active footprint to young adulthood, and there preserve them. While designed to be tested in an area as small as that of a radius of 10 meters, it is designed to be capable of expansion of the active field in increments, as governments permit, until the whole earth is covered simultaneously by the field from one small machine system. One machine could restore the whole earth to Edenic perfection in about 3½ years of machine operating time.

The machine would work by reversing the order to disorder arrow in the second law of thermodynamics. Here are several examples of what that means. Processes with which we are familiar progress in one direction. For examples, fresh picked fruit is fresh only for awhile. Then it ripens too much, spoils, then decays. Newly built barns last quite awhile, but eventually sag, and if never maintained, eventually fall down and collapse. Fresh paint on houses gradually oxidizes, blisters, and peels. A young human grows older, gets wrinkles, loses hair and teeth, gets stooped, gets sick, dies, gets cremated, the ashes are scattered. There are an infinite number of similar possible ordered pathways such as this that could be cited. The events as we have ever observed them always go in one direction. That direction forms one of the important arrows of time. What governs this direction is the order to disorder arrow in the second law of thermodynamics.

The author has discovered a theoretical way to reverse the order to disorder arrow in the second law of thermodynamics. That means everything in the above examples could be played in reverse: Decayed fruit could un-decay, un-spoil, become fresh picked fruit again. Blistered, peeled, oxidized paint could be restored to new again automatically. Scattered human ashes could be automatically re-gathered, unburned, assembled into a dead person, brought to life, and then healed of all diseases and infirmities, made younger, until the process would saturate out at young adulthood. These pathways would all be reversed like this because the order to disorder arrow in the second law of thermodynamics is reversed. This would be possible by the operation of a theorized machine called the Refresher 1. The author has never discovered an existing ordered aging pathway that is not thrillingly better when reversed. Thus the creation and operation of the Refresher 1 is highly desirable—a worthy project at any cost.

This book will describe the restoration that machine could give, summarize the theory of operation, and give the gist of the design of this highly valuable invention.

The inventor at first was so desperate to build this machine that he desired to give away the secrets of how to build it to anyone who desired to do so. But he got absolutely no takers. A short time ago the author received a guaranteed loan solicitation for hundreds of millions of dollars—enough to build and operate the Clean Energy Source and requisite Refresher 1 until the system could produce enough electricity to retire the loan. But the pre-loan fees and permits and working capital needs were in the multiple millions of dollars. There was a need for a wealthy corporate sponsor to advance enough venture capital to cover these necessary pre-loan expenses. But that would not be possible if the trade secrets were given to everyone free.

The trade secrets themselves are costing the inventor millions of dollars consultant fees. The plan now is keeping the trade secrets secret and only giving them to those who need them who adequately pay for them for a license to have and use them to construct and market Clean Energy Sources. Don't worry. The information in this book will not give away the trade secrets. It will disclose only old out of date and insufficient information to construct the inventions. But such old limited information is here disclosed for a couple of reasons—1) to show how long we have been serious in designing and costing this invention; 2) giving the overall gist of the inventions to laymen and government leaders to inspire the masses with the progress in these innovations, and hope for the future.

This book will also present part of a Draft Environmental Impact Statement for the creation and operation of Refresher 1.

CONTENTS

Chapter 1

RESTORATION

The human race is subject to the second law of thermodynamics and all its horrors. All disease, decay, and death are the results of processes working out that law. However, the author has discovered that may not be an inviolable law. It appears the law itself may be reversed in a large or small area. That is, the order to disorder arrow in the second law of thermodynamics may be reversed in a given area, reversing aging, disease, and decay processes.

How can such a thing be? First we must know about the order to disorder arrow in the second law of thermodynamics. And second we must theorize how to reverse it.

Events come in order. A new thermos bottle falls off a countertop and shatters the glass bottle inside. No one expects the thermos to fall back up to the countertop and restore the glass bottle. Why not? Because it is a law for items in a closed system to progress from order to disorder. The whole thermos was more order. The broken thermos is more disorder. From order to disorder is the direction of the order to disorder arrow in the second law of thermodynamics. That arrow is one of several arrows of time which indicate the passage of and direction of time. [1, 2]

The next point to take recognizance of is the nature of the order and disorder in the order to disorder arrow in the second law of thermodynamics. Order requires slightly more energy than disorder. In going from order to disorder, the system loses energy. But here is where the author has something to contribute. He has discovered there are at least two kinds of energy, not just one. He has named them order energy (E_O) and entropy energy (E_S). Order energy is the positive and negative energy in the creation of particles. Entropy energy is the absolute value, term by term, of the order energy in the equations. For instance, when an electron and a positron annihilate, order energy tends to zero. $(E_e) + (-E_e) = 0$. However, entropy energy tends to a positive value. $|E_e| + |-E_e| = 2E_e$. Entropy energy almost always increases. Entropy increases. Entropy energy is the common energy measured every day. Order energy is different. It is the quantum mechanical energy utilized in the creation and annihilation of particles. Entropy energy E_S is involved in the entropy arrow of time. $\Delta E_S \geq 0$. Order energy E_O is involved in the order arrow of time. In the present state of things, $\Delta E_O \leq 0$. The entropy arrow and the order arrow have been thought to be one and the same. But the author has discovered they are two different and separate arrows of time. It is possible to reverse the order arrow without affecting the entropy arrow of time.

How could we reverse the order to disorder arrow in the second law of thermodynamics, and make disorder tend to order?   We would have to find in a closed system a process that would create positive order energy—in other words make $\Delta E_O > 0$. Natural decay processes won't do that. Decay processes create a trace of antimatter in the decay products, which has negative order energy.   The longer or more a system decays, the more negative the change in the order energy ΔE_O. Pair production will not make a net positive change in order energy either. While matter particles produced have positive order energy, the antimatter produced has negative order energy.   The positive and negative order energies cancel out, leaving the net order energy unchanged.   The Standard Model of Physics has no reaction that will yield a positive change of order energy. But the author's model of physics, Electrino Physics, does have one reaction that can take away from the negative order energy and add to the positive order energy, thereby yielding a positive change in order energy. As long as the Standard Model of Physics governs our research, and no other model of physics is permitted to be tested, the human race will be subject to all the horrors of the second law of thermodynamics—aging, disease, decay, and death. But there is hope to reverse the order to disorder arrow in the second law of thermodynamics if a new model of physics can be tested.

The author's model of physics was already presented in an article in the *Galilean Electrodynamics Journal*—"A New Way to Calculate Electron and Muon g/2-factors," in the January-February, 2006 issue. [3] The model is an aether model, where the aether is a sea of bosons of a variety of different spins. In the model symmetric smooth charge distributions cannot have detectable spin. But electrons and muons have detectable spin. Therefore they cannot be symmetric point charges. With the addition of a parsimony postulate, electrons and muons are seen to have two sub particles in them, called semions (half-particles). Semions can fuse to unitons (whole particles). This is the key reaction that can affect the second law of thermodynamics. But electron semions fusing would not produce positive change in order energy. But the fusion of positron anti-semions would produce positive change in order energy—just what we need to reverse the order to disorder arrow in the second law of thermodynamics. These features of the model are carefully and completely derived, amplified, and illustrated in the volume published on the internet, Electrino Physics, by the author. [4] [See now *Electrino Physics* Draft 2, published by Xlibris, and available to order from amazon.com, barnsandnoble.com and 2500 bookstores.]

A positive order energy test facility and complete Electrino Fusion Power Plant [Clean Energy Source] and requisite Refresher 1 have already had a rough design, and has been costed [out of date information]. A leading well qualified builder and agency have been identified. All we lack now is a large corporate sponsor to provide several million dollars pre-loan fees and working capital prior to a senior secured loan of hundreds of millions of dollars for the construction and operation of the Clean Energy Source and its Refresher 1.

[1] Stephen Hawking, *A Brief History of Time—From the Big Bang to Black Holes* (New York: Bantam Books, 1988), pp. 102, 103.

[2] F. Reif, *Statistical Physics*, Berkeley Physics Course—Volume 5 (New York: McGraw-Hill Book Company, 1967), p. 283.

[3] G. L. Ziegler, "A New Way to Calculate Electron and Muon g/2-factors," *Galilean Electrodynamics*, Vol. 17, No. 1, January/February 2006, pp. 11-15.

[4] Gordon L. Ziegler, *Electrino Physics*—from the beginning studies of light to particle structure and a unified field theory (P.O. Box 1162, Olympia, WA 98507 USA: Benevolent Enterprises, last revised November 20, 2008)—available for downloading free at http://www.benevolententerprises.org Book List.

Chapter 2

THEORY

Introduction

The Principal Investigator has discovered a new Grand Unification Theory (GUT). It has deeper symmetry and lower orbital structures than the Standard Model of Physics. It has greater parsimony than the Standard Model. Whereas the Standard Model requires 61 different elementary particles to construct known light and matter [1](page 48), the GUT requires only two different elementary particles to construct known light, matter, and gravitons; and those two different particles can both be ionized from empty space with a single particle.

What differences does this GUT have to the Standard Model? The GUT is an aether model of physics. It has aether special and general relativity, rather than Einstein's aether-less Special and General Relativity. This makes a simple model of gravity and inertia possible [2](Chapter 5). Up until now, uniting special and general relativity in particle physics has been as difficult as uniting fire and ice. This problem is solved with aether special and general relativity in the GUT [2](Chapter 6). Special and general relativistic calculations are both exact fits in the particle structures calculated in this chapter.

The GUT has one postulate that states that cylindrically or spherically symmetric smooth charge distributions cannot have detectable spin. But electrons and positrons have detectable spins. Therefore they must not be symmetric point charges, but have two half charges in them orbiting about each other. The orbiting like charges show that fracton charges come in $\pm e$, $\pm e/2$, $\pm e/4$, and $\pm e/8$ (the Electrino Hypothesis), rather than in $\pm e/3$ and $\pm 2e/3$ (the Quark Hypothesis). The Electrino Hypothesis is very different from the historic and accepted Quark Hypothesis.

11

Yet it does not lead to untenable particle structures. The Principal Investigator has induced the particle structures of all known light, matter, and gravitons, through the simultaneous satisfaction of nine criteria: particle charge, spin, parity, mass, spin feasibility, preceding particles (to avoid duplication), decay schemes, Pauli Exclusion Prinsible, and b-state laws.. [2](Appendix B) Satisfying all the listed Decay Modes published in *Summary Tables of Particle Properties*, by the Particle Data Group [3] (which reference [2](Appendix B) does), is the satisfaction of thousands of tests. Except for being an unknown model, the GUT is in a strong position. Its particle structures are all unique. The quark model particle structures are not all unique.

Another difference of the GUT to the Standard Model is that charged sub-particles of like fusion states can fuse to particles of higher fusion states. [2](Chapter 12) But when they do, they switch from matter to antimatter, or vice versa. Utilizing this phenomenon, the fusion of sub-particles in positrons results in the generation of solely positive order energy (quantum mechanical energy in the creation of particles). This phenomenon is theorized to reverse the order to disorder arrow in the second law of thermodynamics. [2](Chapter 16) (See **Theory**.)

It costs little or no more to build and operate a complete Clean Energy Source with its requisite Refresher 1 than to build merely a scientific test facility to test for electrino fusion reactions. But this project tests three different features of the new GUT: 1) It tests for the fusion of the sub-particles (anti-semions) of positrons into the core particles (unitons) of protons and neutrons. 2) It measures the efficiency of sub-particle fusion for future applications of sub-particle fusion, like power generation. 3) It tests a bonus dividend of the GUT—how to reverse the order to disorder arrow in the second law of thermodynamics, thereby reversing aging, disease, and decay processes.

No present facility anywhere in the world is designed to do these tests. In order to do these tests we must construct a new facility—a positron-positron-collider at over 1876 MeV in the center of mass frame. The spins must be like oriented in the center of mass frame, which means that the spin orientation from one of the accelerators must be reversed relative to that accelerator. The spins must be axial. Such a collider would test for sub-particle (electrino) fusion, which would be a test of the Electrino Hypothesis (fractons come in $\pm$ e, $\pm$ e/2, $\pm$ e/4, and $\pm$ e/8) as opposed to the Quark Hypothesis (fractons come in $\pm$ e/3 and $\pm$ 2e/3). Ratios of protons-neutrons produced relative to positrons accelerated would directly measure the efficiency of the sub-particle fusion. Another indirect test for the efficiency of the fusion process would be the radius of the affected area in the reversal of the order to disorder arrow in the second law of thermodynamics. Strangely enough, the affected radius is inversely proportional to the positron beam current in theory, with 1 pA corresponding with 96 (roughly 100) m affected radius. Thus the second law of thermodynamics aspect of the experiment is a valuable adjunct to the study, and not just a humanitarian spin-off to the project.

The second law effect, to be perceivable and useful, requires very low beam currents. But low fusion rates could demonstrate sub-particle fusion and efficiencies. So low beam currents would work for all features of the GUT to be tested by this facility. Therefore the proposed facility is designed to run on very low beam currents.

What if the facility works as designed? Within the affected area great changes would take place in people and their environs. Aging of man, animals, and vegetation would rapidly be reversed, and the order increased to maximum order—young adulthood. Diseases would be backed out of existence. Once these properties were demonstrated, you know great crowds would flock to the facility's property for healing and age reversal. We must think ahead and provide them with attractive park grounds and restrooms. Thus, such is part of this proposal.

The applicant proposes to carry out this project mostly by employing qualified, competitive sub-contractors to do the work. The principal qualified sub-contractor is James M. Potter, Ph.D. of JP Accelerator Works, Inc. of Los Alamos, New Mexico.

The significances of the project are several. The first unique significance of the project is that quantities will be tested that have never been tested by man before. The highest energy electrons that have been collided were 700 MeV. Never have they been collided at 939 MeV or more. That is simply because the Standard Model of Physics predicts nothing interesting for collisions of those energies, other than collisions and recoils. But that is a great mistake to neglect to test those energies, for collisions of 939 MeV or more for electrons or positrons yield fusion of sub-particles and the switch of matter to antimatter (or vice versa), according to the Electrino Fusion Model of Elementary Particles.

A second significance of the Electrino Model tests to be conducted at this proposed facility is particles would be fused and switched from matter to antimatter or vice versa, which has never knowingly been observed before, It would be of greatest interest in the scientific community.

A third significance of the proposed tests is that by these means the human race can learn how to reverse the order to disorder arrow in the second law of thermodynamics, thereby reversing aging, disease, and decay processes. Contemplated and understood, that is a highly desirable result. It would revolutionize human existence.

A fourth significance of this project is that one small facility as proposed can do so much good in the world. It can test for sub-particle or electrino fusion. It can test for the efficiency of electrino fusion for power generation purposes. It can reverse aging, disease, and decay processes in a small test area for test purposes. It can reverse aging, disease, and decay processes in larger and larger areas, as the voting public and governments permit, until the whole earth is affected—from a single small facility. This facility is highly cost effective and desirous.

This project is related to the long term goals of the Principal Investigator—which are to eliminate death, disease, old age, and pain from this planet. This would be one way of doing this. Even if the facility failed to do this, another long term goal of the Principal Investigator would be advanced—the progress of truth and science. No matter how these tests turn out, more would be known, and giant holes in the data would be filled.

Present knowledge in this field is nonexistent. This project would greatly advance knowledge in this field.

Theory

An aether model of relativity more easily accounts for all the tests of relativity than Einstein's aetherless Special and General Relativity. [4] An aether model of relativity also easily accounts for gravity and inertia. [2](Chapter 5) An aether model of relativity harmonizes the conflicting claims of Special and General Relativity in particle structure, provided fracton charges come in $\pm$ e, $\pm$ e/2, $\pm$ e/4, and $\pm$ e/8 (the Electrino Hypothesis) rather than in $\pm$ e/3 and $\pm$ 2e/3 (the Quark Hypothesis). [2](Chapter 6) The Principle Investigator has induced the structure of all known particles using the Electrino Hypothesis. [2](Appendix B) The result is an alternative system of particle structures to the Quark Model. The electrino particle formulae are all unique. Not so with quark particle formulae.

Another thing the Electrino Model does, that the Quark Model does not do, is provide for electrino fusion. Four of a fusion state may fuse to two of the next higher fusion state. [2](p. 197); [2](Chapters 12, 15, and 16) According to the induced particle structures referred to above, electrino fusion makes the particles switch from matter to antimatter, and vice versa, and also affects the second law of thermodynamics in a calculable radius. We will treat this latter aspect with greater detail.

The human race is subject to the second law of thermodynamics and all its horrors. All disease, decay, and death are the results of processes working out that law. However, the author has discovered that may not be an inviolable law. It appears the law itself may be reversed in a large or small area. That is, the order to disorder arrow in the second law of thermodynamics may be reversed in a given area, reversing aging, disease, and decay processes.

How can such a thing be? First we must know about the order to disorder arrow in the second law of thermodynamics. And second we must theorize how to reverse it.

Events come in order. A new thermos bottle falls off a countertop and shatters the glass bottle inside. No one expects the thermos to fall back up to the countertop and restore the glass bottle. Why not? Because it is a law for items in a closed system to progress from order to disorder. The whole thermos was more order. The broken thermos is more disorder. From order to disorder is the direction of the order to disorder arrow in the second law of thermodynamics. That arrow is one of several arrows of time which indicate the passage of and direction of time. [5, 6]

The next point to take recognizance of is the nature of the order and disorder in the order to disorder arrow in the second law of thermodynamics. Order requires slightly more energy than disorder. In going from order to disorder, the system loses energy. But here is where the author has something to contribute. He has discovered there are at least two kinds of energy, not just one. He has named them order energy (E_O) and entropy energy (E_S). Order energy is the positive and negative energy in the creation of particles. Entropy energy is the absolute value, term by term, of the order energy in the equations. For instance, when an electron and a positron annihilate, order energy tends to zero. $(E_e) + (-E_e) = 0$. However, entropy energy tends to a positive value. $|E_e| + |-E_e| = 2E_e$. Entropy energy almost always increases.   Entropy increases. Entropy energy is the common energy measured every day. Order energy is different. It is the quantum mechanical energy utilized in the creation and annihilation of particles. Entropy energy E_S is involved in the entropy arrow of time. $\Delta E_S \geq 0$. Order energy E_O is involved in the order arrow of time. In the present state of things, $\Delta E_O \leq 0$. The entropy arrow and the order arrow have been thought to be one and the same.   But the author has discovered they are two different and separate arrows of time. It is possible to reverse the order arrow without affecting the entropy arrow of time.

How could we reverse the order to disorder arrow in the second law of thermodynamics, and make disorder tend to order? We would have to find in a closed system a process that would create positive order energy—in other words make $\Delta E_O > 0$. Natural decay processes won't do that. Decay processes of particles create a trace of antimatter in the decay products, which has negative order energy. The longer or more a system decays, the more negative the change in the order energy ΔE_O. Pair production will not make a net positive change in order energy either. While matter particles produced have positive order energy, the antimatter produced has negative order energy. The positive and negative order energies cancel out, leaving the net order energy unchanged. The Standard Model of Physics has no reaction that will yield a positive change of order energy. But the author's model of physics, Electrino Physics, does have one reaction that can take away from the negative order energy and add to the positive order energy, thereby yielding a positive change in order energy.

As long as the Standard Model of Physics governs our research, and no other model of physics is permitted to be tested, the human race will be subject to all the horrors of the second law of thermodynamics—aging, disease, decay, and death. But there is hope to reverse the order to disorder arrow in the second law of thermodynamics if a new model of physics can be tested.

The author's model of physics was presented in brief in an article in "A New Way to Calculate Electron and Muon g/2-factors," in the January-February, 2006 issue of *Galilean Electrodynamics*. [7] The model is an aether model, where the aether is a sea of bosons of a variety of different spins. In the model, symmetric smooth charge distributions cannot have detectable spin. But electrons and muons have detectable spin. Therefore they cannot be symmetric point charges. With the addition of a parsimony postulate, electrons and muons are seen to have two sub particles in them, called semions. Semions can fuse to unitons (whole particles). This is the key reaction that can affect the second law of thermodynamics. But electron semions fusing would not produce positive change in order energy. But the fusion of positron anti-semions would produce positive change in order energy—just what we need to reverse the order to disorder arrow in the second law of thermodynamics. These features of the model are carefully and completely derived, amplified, and illustrated in *Electrino Physics*, by the author (published on the Internet since August 15, 2007 at http://benevolententerprises.org Book List. The work is downloadable for free at that site.). [2]

Positron sub-particle fusion might not only take the electrinos from disorder to order. It could make other physical processes in a local area go from disorder to order. The positron fusion not only violates the second law of thermodynamics, it reverses the order to disorder arrow of that law in a local area, making other processes in that area reverse. Let us consider that process more to see how it might be regulated.

We guess the desired relationships for reversing the order to disorder arrow in the second law of thermodynamics through dimensional analysis. We want to solve for r, the maximum radius in which the reversed law would be effective. There is a way we can obtain a length from combinations of our variables and constants. That way is in the right hand side of Eq. (1). The whole expression is the thermodynamic relation we are seeking. The thermodynamic relation is:

$$(\Delta E_o)_t > 0 \ where \ r < \frac{(\Delta E_o)_1 \ c}{ik}, \tag{1}$$

where E_o is the order energy–the positive or negative energy in the pair production of particles; ΔE_o is the change in the order energy, where $(\Delta E_o)_t$ is the change in the total order energy of the system, and where $(\Delta E_o)_1$ is the change in the order energy for a single source reaction—for a positron fusion reaction it is approximately 2×10^9 eV/collision $\times 1.6 \times 10^{-19}$ joules/eV $= 3.2 \times 10^{-10}$ joules/collision; c is the speed of light—approximately 3.0×10^8 m/s; we shall solve for the effected radius r; i is the beam current in each beam in Coulombs per second (we will solve for 10^{-11}); k is the ratio of particle energy to particle charge. This energy per charge is the accelerated energy of the particle (roughly 1×10^9 ev times 1.6×10^{-19} joules/ev $= 1.6 \times 10^{-10}$ joules) divided by the charge of each positron ($q = 1.6 \times 10^{-19}$ coulombs), which equals 10^9 joules per coulomb. The collision efficiency eff is not needed in this equation, because the result is not in particles, but is already in collisions.

Incredibly, the lower the current, the bigger the radius of the affected area. And the greater the current, the smaller the radius of the effected area. With 10^{-11} A beam currents, the effected radius r solves for 9.6 meters—roughly 10 meters, which describes a small area—less than a tenth of an acre.

To get an idea of the positron beam currents needed to reverse the order to disorder arrow of the second law of thermodynamics in what size of affected radius, see Table 1 below.

For an area the size of	r	beam current
House	10 m	10 pA
four football fields	100 m	1 pA
community	1 km	100 fA
city	10 km	10 fA
Israel	160 km	0.6 fA
U.S.	2,400 km	0.04 fA
World	13,000 km	0.008 fA
Sun	1.7E11 m	6E-22 A

Table 1. Beam currents versus affected radius for reversal of the order to disorder arrow of the second law of thermodynamics.

[1]   David Griffiths, *Introduction to Elementary Particles (New York: John Wiley &Sons, Inc., 1987).*

[2] Gordon L. Ziegler, *Electrino Physics* (P.O. Box 1162, Olympia, WA 98507-1162 USA: by author, last revised September 10, 2007)[downloadable for free at http://benevolententerprises.org Book List]. [See rather *Electrino Physics* Draft 2 (last edited May 1, 2014) at xlibris.com, amazon.com, barnsandnoble.com or 2500 bookstores.]

[3] C. Caso, *et. al.* (Particle Data Group), "Summary Tables of Particle Properties," (including "Gauge and Higgs Bosons Summary Table," "Lepton Summary Table," "Meson Summary Table," and "Baryon Summary Table), *CRC Handbook of Chemistry and Physics*, 80th Edition, David R. Lide, Ph.D., Editor-in-Chief (Boca Raton: CRC Press, 1999-2000), pp. **11**-1 to **11**-42.

[4] Tom Van Flandern, Univ. of Maryland and Meta Research, *Open Questions in Relativistic Physics*, edited by Franco Selleri (Montreal: Apeiron, 1998), pp. 81-90, as quoted by "What the Global Positioning System Tells Us About Relativity," Meta Research, http://www.metaresearch.org/cosmology/gps-relativity.asp.

[5] Stephen Hawking, *A Brief History of Time*—From the Big Bang to Black Holes (New York: Bantam Books, 1988), pp. 102, 103.

[6]   F. Reif, *Statistical Physics*, Berkeley Physics Course—Volume 5 (New York: McGraw-Hill Book Company, 1967), p. 283   .

[7] G. L. Ziegler, "A New Way to Calculate Electron and Muon g/2-factors," *Galilean Electrodynamics*, January/February 2006, Vol. 17, No. 1, pp. 11-15.

Chapter 3

Old Outdated Refresher 1 Design Specifications

Size of accelerator (folded)	20 m long by 3 m wide
Diameter of accelerator	100 mm (plus cooling channels)
Beam aperture	7 to 10 mm
Type of accelerator	Folded linear accelerator with pulsed klystron rf power supplies and S-band cavities (2856 MHz)
rf power supplies	Eight 35 to 50 MW pulsed klystrons
duty factor	0.1% (peak current 1000 times average current)
Average power	400 kW (20 kW per meter of accelerator)
klystron efficiency	~50%
total system power	800 kW
cooling water requirement for each 5 m section	5 to 10 gpm
cooling water required by each klystron	~ 5 gpm
cooling towers capacity	800 kW

Creation time total (if not
super funded ($50 million))   3 years
  Design time (beam dynamics,
  rf power systems, cooling,
  and computer control)   1 year

  Fabrication and subassembly
  testing   18 months

  Installation and commissioning   6 months

Creation time total (if super
funded ($100 million)   1 year

Chapter 4

OLD OUTDATED COST ANALYSIS

As of October, 2007, the cost for the accelerator equipment constructed in 3 years is $38 million. The total cost of that project would be about $50 million.

As of October, 2007, the cost for the accelerator equipment constructed in 1 year is $93 million. The total cost of that project would be about $100 million

These costs change with time, going up an average of 3% per year. A current detailed cost analysis is available on request if you have a need to know. Contact

Gordon L. Ziegler
P.O. Box 1162
Olympia, WA 98507-1162 USA
ben_ent100@msn.com.

Chapter 5

Clean Energy Source and Refresher 1

I have discovered a new Grand Unification Theory (GUT) which unites all the forces (which Einstein tried to do for thirty years and failed to do)—*Electrino Physics*, Chapters 7 and 8 and *The Physics of Genesis*, Chapter 3. Scientists have sought these truths so hard for so long that someone questioned if and when someone did discover these, would it bring to light any additional powers to man, or would the ultimate physical truth be just some quaint curious mathematical formula. I am here to tell you that this precious light is not a quaint curiosity. It really does open the door for the most revolutionary benefits to man— among them the Clean Energy Source and Refresher 1.

This new light on physics has two problems: One problem is that traditional physics has taken some wrong turns, and is entrenched in some serious falsehoods that need to be corrected. The new theory needs to be experimentally tested. But all grant applications are reviewed according to the Standard Model of Physics. All Grant Applications are rejected that do not agree with the Standard Model of Physics in every particular. The three main instances of that are 1) Einstein's Aether-less Special Theory of Relativity in which Einstein claimed that nothing could go faster than the speed of light. It is impossible to unite all the forces or have a Clean Energy Source or a Refresher 1 unless certain classes of particles travel faster than the speed of light. 2) Another problem is P. A. M. Dirac's doctrine that electrons are spinning point charges. They are not. But if that is explained in a Grant Application, the Application is rejected. 3) By far the worst instance of doctrinal error entrenched in physics is Murray Gel Mann's Quark Hypothesis that the electric particle fractons come in $\pm 2e/3$ and $\pm e/3$. They do not. They come in $\pm e/8$, $\pm e/4$,

± e/2 and ± e, according to the author's Electrino Hypothesis. See *The Messiah's Inventions,* Section 3, the three chapters starting with the word "impossible."

The other problem to the acceptance of this light is the centuries' long great dichotomy between science and miracles. Scientists claim not to believe in miracles, but in natural law. When the new GUT explains how to do undeniable miracles through well explained new science, it is way too good to be true! and it is rejected out of hand, with no money allocated to those projects! Children can be taught it in simple terms, but it absolutely confounds the Ph.D professors. Did not the laboring and speculating scientists want new powers from their research? What if correcting three errors in the Standard Model of Physics unleashes a plethora of miraculous powers? Bonanza! Shouldn't the scientists and government leaders be thankful, not rejecters? If it only took a few million dollars to test so many miracle powers, wouldn't this light be way too good to be false?

Well, what can your GUT do for the energy crisis and global warming? It can make a relatively inexpensive, carbon emission-free, radioactive waste-free electric power reactor, in one sense 1000 times as efficient as nuclear power—needing refueling only once every 100-200 years, and then without hazard, just substituting in the particle collider a prefabricated part. The Clean Energy Source does not need hazardous radioactive fuels or Hydrogen. It can annihilate copper or brass for fuel. Annihilating one old Penney in the collider would produce up to $1,872,000.00 of electricity at $0.03/kwh. Depending on the net efficiency of the Clean Energy Source (to be determined in the micro-design phase of this project) the new equipment can pay for itself with $0.03/kwh electricity in five years or as little as one year operational time.

But none of that is possible without the building and operating of another related machine. The Energy Source would not be efficient enough to be self-sustaining without the operation of the Refresher 1. Neither would it be a clean source. It would have radioactive wastes without the Refresher 1. We want a clean, efficient energy source. There is even legislation regarding that! But who can handle a Refresher? It is altogether too miraculous for most! What would it do?

Miraculous Effects of the Refresher

Reverse aging for adults

The simplest effect of the Refresher to understand is reversing adult aging. Old people can be made young adults again in the active footprint of the Refresher. This effect for positron anti-semion fusion does not really back up time or the clock. It merely reverses the order to disorder arrow in adults. It saturates at the maximum state of order—which is young adulthood. It reverses adult aging at a rate of about 1836 times as fast as the rate the original adult aging occurred. A century of aging can be reversed in just under 20 days.

Resurrections from the dead

The reverse aging occurs also for bodily remains—re-assembling dust and bones into living beings again. All the dead of all ages of earth's history would be resurrected in about 3½ years of Refresher machine time, starting with those who died most recently.

Backing diseases out of existence

In the process of reverse aging, diseases would be backed out of existence. This would work also for difficult diseases like HIV AIDS, cancer, and cystic fibrosis.

Reversing all decay

Spoiled fruit would un-spoil in the active footprint of the Refresher. Fresh fruit would stay at the maximum state of order for fruit forever—fresh picked fruit. And this would be without refrigeration. This would amount to a new kind of food preservation without canning or freezing.

This process would un-decay everything in the Refresher footprint, not just fruit. And the footprint could be enlarged to cover the entire earth.

Reversing pollution out of existence

In the Refresher footprint, all pollution would be reversed out of existence. Depending on the Refresher control settings, this effect could be world-wide.

"Raising up the foundations of many generations" Isaiah 58:12.

The Refresher would automatically rebuild previous decayed structures. It would rebuild and restore the entire earth.

Reversing forest fires

The Refresher not only would stop forest fires in its footprint, but would reverse the fires—restoring all that was lost—animate and inanimate, including lost trees and homes.

Reversing all calamities;
Reversing all effects of war;
Preventing all munitions from firing;
"Making wars to cease to the end of the earth." Psalm 46:9.
Removing sinful propensities from people, including criminals;
Emptying prison houses;
Making possible and efficient Clean Energy Sources.

The blessings of the Refresher are endless. In short it would restore earth to Edenic perfection in about 3½ years of machine time.

Can people handle all that?

Gordon Ziegler DBA
Electrino Energy
4401 37th Ave SE Unit 17
Lacey, WA 98503-3576
Email: ben_ent100@msn.com

1.4 Facility Environmental Impacts and Mitigations

The facility (Refresher 1) is a low beam current, high energy (over 1876 MeV in the Center of Mass Frame) positron-positron accelerator-collider. One environmental impact is that the system will emit positron, beta, and gamma radiation. Because the beam currents are very low (less than or equal to 10 pA each beam), the radiation levels (yet to be determined) of the accelerators and collider will be very low. The radiation levels will be mitigated further by twelve feet of earth shielding. The resultant residual radiation levels will be far below the federal de minimus level for radiation exposure to the general public (0.01 mSv per year to members of the public).

Whereas the general public will not be exposed to significant amounts of radiation by this facility, the general public may be exposed to the active field in a small controlled test area with human use protocols and human use consent forms according to federal and Western Institutional Review Board (WIRB) standards. After 10,000 individuals have been successfully treated without any negative side effects occurring, Refresher 1 will seek government authority to expand the operation to areas outside of the test boundaries (five acres) so that humans can freely expose themselves to a field caused by the reversal of the order to disorder arrow in the second law of thermodynamics in the grounds outside of this facility. The field is beneficial to individuals and/or the planet in every way, so mitigations will not be necessary for the active field, except several relevant advisories will be posted for the visitors' comfort and pleasure. The following is the notice in English that will be posted around the periphery of the active area:

"NOTICE:

- **Welcome to the Eden-like field of Refresher 1. There is no charge for this blessing. But we accept donations.**

- **You may find that your aging is reversed to young adulthood in the active area of this field. You may also find all your diseases backed out of existence, so you find yourself in perfect health.**

- **However, there are certain things about this field you should be aware of before you enter the active area: the field will try to heal all your old wounds. All body pierced jewelry should be removed before entering.**

- **The field also will attempt to resurrect dead animals and the leaves, stems, and roots of vegetables. To avoid the discomfort of animals and vegetables re-organizing** in your stomach, it is recommended that visitors eat only fruits, nuts, grains, legumes, and fruit-like vegetables in the active area and 72 hours before entering the active area. Food designed for Eden-like living will be provided for no charge for those within the active area and those waiting their turn to enter."

[Refresher 1 is not only for good health. Particle detection equipment can be placed in the containment building of the accelerators to test for revolutionary physics data.]

2.0 Purpose and Need of Proposed Action

The purpose for this proposed action is three-fold: 1) detect high energy sub-particle fusion; 2) measure the efficiency of sub-particle fusion for power generation purposes; and 3) test the reversal of the order to disorder arrow in the second law of thermodynamics, thereby reversing aging, disease, and decay processes in a test area, and in larger and larger areas as the public demands.

The need of the proposed action is also three-fold: 1) high energy sub-particle fusion is needed to simplify, give added symmetry and an additional layer of orbits to particle structure and increase parsimony to the science; 2) e^- e^+ collisions have 1.602×10^{-19} efficiency. But e^+ e^+ and e^- e^- collisions have a theoretical efficiency of 1.0 in the new model at over 1876 MeV in the Center of Mass Frame, due to the magnetic and weak forces making the particles smart bombs with respect to the strong force. That is a big difference of efficiency. Should all the efficiencies be 1.602×10^{-19}? If so, a new breed of power generators would not be possible which would be very helpful. Our facility would be able to measure the efficiency in more than one way. 3) Every sickness, every pain, and every death on this planet is a need for reversing the order to disorder arrow in the second law of thermodynamics, which our proposed Refresher 1 Facility might be able to do in a small test area, in larger and larger areas if the public demanded and governments permit.

3.0 Significant Issues or Sensitive Receptors

The Refresher 1, reversing the order to disorder arrow in the second law of thermodynamics, does many things that seem too good to be true. Thus the reviewer of these things may reject this project out of hand. But all these things happen naturally when that arrow of time is reversed. Notice a possible sequence under the existing second law of thermodynamics:

Healthy young adult → aging → cancer → death → cremation → scattering of ashes.

What if the arrows were reversed—all of the arrows? Well, not only could one be healed from cancer, made a healthy young adult again, but scattered ashes could gather themselves again, a person could be un-cremated, come to life again, be healed of cancer, made young again, and made a healthy young adult again, which is the maximum state of order in human beings. The system goes to and is stable at the maximum state of order with the reversed order to disorder arrow. But notice in all this, it is order that is increased, not the clock that is turned backward. The clock continues forward. Sometimes it is difficult to keep that distinction in mind in imagining second law reversals.

Let us imagine some second law reversals: A fallen down barn with weathered wood and rusty nails falls back upright, ceases to sag, and becomes a new barn again with new wood and new nails. A burned out forest un-burns. The burned to death forest creatures come alive again and are restored to health. In the battle field, explosives refuse to explode. Killings stop. Exploded ordinance un-explodes, restoring all that was lost. Slain soldiers are resurrected. Maimed soldiers are made whole again. There is no more pain or sorrow. Enemies are grateful for the blessings of life, and are reconciled. Generation after generation is resurrected, and the foundations of many generations are raised up. Such possible scenarios are endless. They all seem too good to be true, but they would all be possible if only the order to disorder arrow in the second law of thermodynamics could be reversed in a controllable area. That may be the case with the proposed Refresher 1. In view of the incredible blessings it may offer to us, we must permit the Refresher 1 to be built and tested.

What will be impacted the most? Those who have been dead the longest, whose remains are the most scattered. It would take the most machine time to resurrect them. It would take nearly three and a half years machine time to resurrect those dead 6,000 years.

What are the most important impacts? The elimination of pain, suffering, disease, decay, and death. The Refresher 1 will be designed to do this in a small test area, in larger and larger areas as the public demands and governments permit, and for the entire earth at once, when all the governments of the world OK it. [The planet would not be over crowded. See Section 1.4.]

It would be desirable to build one or two more Refreshers as backups when there is an outage.

16

Draft Environmental Impact Statement: High Energy Sub-particle Fusion Test Facility

4.0 Alternative Development

Are there better ways to meet the Purpose & Need? No. The only alternative the human race has had for thousands of years is the religious promise that God would come in the end of the world and do this for the saved few, whereas the lost great majority would perish eternally. The Refresher 1 promises to save everybody, without the loss of one. There is no scientific or religious alternative to this that can offer such a positive promise. And we do not have to wait in faith for centuries or thousands of years. We can have it as soon as we can build and test the device—in one year.

5.0 Affected Environment – Existing Conditions

The environment and wildlife habitats are increasingly stressed, depleted and threatened by increasing and intensifying natural disasters and over harvesting by man. Hundreds of species are going extinct. The planet is experiencing the dangers and loss of habitat by global warming. The environment still supports life, but time is apparently running out.

The Refresher 1 can operate in a small test area, but it is designed to operate globally. No baseline environmental data is provided for its small test area, but every large facility in the world, such as nuclear reactors, can provide current environmental data of their areas, which can be baseline environmental data for Refresher 1 gone global.

6.0 Environmental Consequences
Prediction of Environmental Impacts

The Environmental Consequences of the field of Refresher 1 have already begun to be told in the Notice posted and given out to visitors entering the field, and in section **3.0 Significant Issues or Sensitive Receptors**. It is impossible to predict every occurrence of law reversal that would be experienced. However additional typical cases that would be representative are presented here for both positive and negative impacts: Aging concrete would become like new concrete again. Houses with blistered paint on siding boards would not only have the paint become like new, but the underlying boards become like new—and there remain stable. Oxidized paint on cars would un-oxidize and become like new. The worn parts would un-wear. The engines would not operate backwards (that would be the backing up of the clocks, not the reversal of the order to disorder arrow).

There are considerations on a global scale that would not be a problem in a small test area. Planes and rockets would fly forward, but would be reversed in aging. Their parts would not wear out or fail. If the decision were made to go global, earthquakes would happen backwards. But instead of causing serious damage, the rubble would be convulsed and fused back into the original beauty of the buildings.

If things went to maximum order, people, animals, and vegetation would glow with coherent light suitable for individual flight like angels, not needing rockets or planes. There would be a bonanza of energy efficiency and availability. With so much light, the sun would not be needed.

7.0 Mitigation – What Will Be Done to Reduce or Prevent Impacts?

The field of Refresher 1 has impacts at every turn. But most or all of them are highly positive, not negative. They describe a much better world. It is the opinion of the Principal Investigator, Gordon L. Ziegler, that there are no true negative impacts from the reversal of the order to disorder arrow in the second law of thermodynamics.

It would be very frightening to the non-initiated, however, to see faces glow when it was not expected. Thus the best mitigation of the impacts would be an educational program through the media explaining these things. That work starts here with this DEIS.

www.ingramcontent.com/pod-product-compliance
Lightning Source LLC
Chambersburg PA
CBHW021049180526
45163CB00005B/2352